Competitive Intelligence

Fast, Cheap & Ethical

An essential guide for managers, start-ups, entrepreneurs & innovators

Rob Duncan

authorHOUSE®

AuthorHouse™
1663 Liberty Drive, Suite 200
Bloomington, IN 47403
www.authorhouse.com
Phone: 1-800-839-8640

© 2008 Rob Duncan. All rights reserved.

No part of this book may be reproduced, stored in a retrieval system, or transmitted by any means without the written permission of the author.

First published by AuthorHouse 6/25/2008

ISBN: 978-1-4343-0641-8 (sc)

Printed in the United States of America
Bloomington, Indiana

This book is printed on acid-free paper.

Foreword

When asked to write a foreword to Rob Duncan's book, I was honored by the request but also struck by how similar the world of competitive intelligence is to the tradecraft of intelligence analysis as practiced by law enforcement agencies, national security and regulatory bodies. In many ways, they are flip sides of the same analytic coin. While the subject matter differs, both practitioners utilize the hard currency of intelligence for strategic advantage, and employ similar tools and techniques in the quest for information supremacy. Whether it's the financial bottom line or catching bad guys, actionable intelligence trumps raw information and Duncan shows us how it's done in the business sector.

The tools, techniques and lessons presented in *Competitive Intelligence: Fast, Cheap & Ethical* help both beginner and expert make sense of the daily stream of data, thereby transforming raw information into intelligence for strategic profit. Duncan encourages us to go beyond traditional sources of data by employing all of our five senses in assembling competitive intelligence, thus letting us access a rich tapestry of information. He praises the virtues of predictive analytics, the holy grail of all intelligence, and urges us to keep it simple. These are sage words of advice for both competitive intelligence professionals and intelligence analysts.

Competitive Intelligence: Fast, Cheap & Ethical is a timely compilation of wisdom from Duncan's extensive and accessible journey into the field of competitive intelligence. It demystifies CI tools and techniques, often perceived as arcane and unobtainable, putting them within reach of practitioners, students and observers of this exciting area. Duncan is to be commended for his thoughtfully-presented and lively treatment of this important topic. It belongs on the bookshelf of every serious analyst.

Alex Tyakoff, MA
Delta Police Department

Contents

Part One: **Core CI Skills** 5
Chapter 1: Curiosity 9
Chapter 2: Sensory Awareness 13
Chapter 3: Numeracy 23
Chapter 4: Research Ability 27
Chapter 5: Ethics 31

Part Two: **Applied CI Skills** 37
Chapter 6: Industry Analysis 39
Chapter 7: Internet Competitor Analysis 49
Chapter 8: Qualitative Market Research 61
Chapter 9: Quantitative Market Research 75
Chapter 10: Mystery Shopping 81
Chapter 11: Human Intelligence 89
Chapter 12: Trade Show Intelligence 103
Chapter 13: Executive Profiling 109

Part Three: **Tying It All Together** 115
Chapter 14: CI as a Daily Practice 117

Who Needs This Book?

"Agility means that you are faster than your competition. Agile time frames are measured in weeks and months, not years." Michael Hugos

If you are thinking about starting your own business, already own a home-based, small or medium-sized business, are a manager in a company of any size, or are an innovator, you need this book. To make decisions about your business, you need intelligence. Not the kind of intelligence measured with IQ tests, but the kind that involves your ability to find, interpret and respond to the information around you. It is vital for entrepreneurs and managers to be able to understand what their competitors are doing, how they think and what they are likely to do in the future.

Despite being in increasingly competitive environments, relatively few entrepreneurs or companies practice meaningful levels of competitive intelligence or CI. This may be because of misperceptions about what CI is. Many perceive CI as corporate espionage, as too time-consuming and expensive, or even as illegal. This book focuses on presenting CI in a realistic light, and as something that can be easy, inexpensive and ethical.

This book gives you a number of tools to help keep tabs on your competitors and predict how they will behave in the future. Best of all, you can follow all of the techniques presented in this book at virtually no cost and in a legal and ethical manner.

The subtitle of this book, *Fast, Cheap & Ethical*, reflects my belief that fast intelligence is better than perfect intelligence. The winning company is the one that solves part of a problem in a few minutes and begins developing counterattack strategies immediately, while filling in their intelligence on the fly. Waiting for several weeks to get an exhaustive answer is not a winning option in highly competitive environments.

Similarly, the book's subtitle reflects my belief that good CI need not be expensive. It's possible to spend a fortune on computerized dedicated CI systems, or to attempt to harness large enterprise systems in service of CI. That's fine, but the typical manager or entrepreneur can do top-notch CI without spending much money at all. Once ingrained as a set of instincts and behaviors, CI becomes something you can do all the time, often from your desktop with minimal financial investment.

Finally, the subtitle of the book reflects a profound conviction that the best CI is ethical CI. The most satisfying challenge in CI is to gather the very best intelligence without lying, cheating, breaking laws or ethical boundaries. To gain an edge on a level playing field, while observing the highest ethical standards, is what the game is all about. To cut corners is not just unethical, it's mentally lazy.

The book is divided into three sections. Chapters 1 through 5 present some core skills you'll need to be a good CI practitioner. Chapters 6 through 13 build on those core skills with a set of applied intelligence techniques. Finally, Chapter 14 presents some suggestions for creating a system that will help you get the most out of your company's CI efforts.

As a disclaimer, let me say few of the CI techniques presented in this book are uniquely mine, but have been gleaned from years of reading, sharing ideas with practitioners and thought leaders, attending seminars and conferences and teaching CI at the college level. However, the practice exercises, the arrangement of the subjects in the book, as well as the focus on core skills, are my own.

The intention behind this book is to popularize these techniques and put them in the hands of entrepreneurs and managers who can make use of them. I gratefully acknowledge the many mentors, teachers, colleagues and influences I've had along the course of my development in CI.

Rob Duncan

PART ONE:
Core CI Skills

"Intelligence is quickness in seeing things as they are." George Santayana

Competitive Intelligence Defined

In the very broadest sense, intelligence is anything that informs your ability to make decisions. Not merely hard facts and data, but also observations, guesses and all the other forms of human sensory, analytical and intuitive experience. Annual reports, website information, satellite imagery, overheard conversations, survey results, focus group discussions, store visits, speculation and hunches are all forms of intelligence.

Competitive intelligence then, is the harnessing of all forms of intelligence you can use to make decisions, with the overall goal of gaining sustainable strategic advantage.

In short, if you can get even one percent ahead of the competition through your intelligence activities, you may be able to gain a sustainable competitive edge that cannot be overcome. If you can do this quickly, affordably and ethically, you'll become a better and more nimble strategist than your competition.

CI Past, Present and Future

The ideal use of CI is to predict future actions, yet of the firms that do CI, the vast majority focus on the past which provides little of predictive value. In the pre-Internet era, CI often consisted of gathering the competitors' annual reports and sifting through them to see "how they did." There is no doubt some value in examining a competitor's past performance, especially in terms of financial indicators, cost structures and stated goals and plans, but this type of backward-looking CI should only be the starting point. Predicting the future actions of competitors is where the ultimate strategic advantage can be realized.

The next level up in sophistication is tracking what the competition is doing currently. The Internet makes this fairly easy and most smart companies are tracking changes in competitors' website information, scanning for news headlines and staying alert to job postings that can be mined for future plans. Most companies also make use of surveys and focus groups to gain insight into consumer perceptions about their company and the competition. Some companies also undertake formal or informal mystery shopping activities to keep track of pricing, service levels and other competitive variables.

This information is critical to effective decision making, and when combined with historical CI, provides a potent tool for gaining strategic advantage. The problem is that you are still only dealing with what has happened in the past and what is happening in the present moment. The decisions that led to the competitor's current behavior were made long ago. What if you could have predicted these decisions? What if you could have predicted them before the competitor made them? Here lies true strategic advantage and this is the highest goal that CI can aspire to.

Predictive CI builds upon your knowledge of the competition's past and present actions, in order to predict the future. Beginning with an in-depth understanding of your own company and your industry, you then use intelligence to predict how the competition will act in the future. That is the key thrust of this book. If you can better predict how competitors will react to future events, you can gain a sustainable strategic advantage.

You only need to be one percent better than your rivals to sustain endless competitive advantage. Predictive CI relies on an in-depth understanding of what exactly drives competitive advantage in your industry, as well as what capacities, constraints and personalities drive actions and decisions in your rival companies. To develop that depth of understanding, you must first acquire some core CI skills.

Core CI Skills

To be a good CI practitioner, there are a number of tools and techniques you can learn. Underpinning these are several core skills that you need to develop or harness.

In the following chapters, you'll explore a number of these skills, including curiosity, sensory awareness, numeracy, research ability and ethics. These core skills are well developed in a top-notch CI analyst, but nobody is born with a full set of them. Each person doing CI will have some areas that come naturally and some that need to be developed. To help hone your skills, each chapter includes easy, fun exercises for you to try.

1

Curiosity

"Necessity may be the mother of invention, but curiosity is the mother of discovery." Charles Handy

Key Chapter Points:

- ❖ Curiosity is the foundation of all good competitive intelligence
- ❖ Methods to improve day-to-day curiosity

Understanding People's Behavior and Motivation

The exchange of goods, services and money causes people to behave in a variety of interesting, challenging, perplexing and frustrating ways. Understanding these behaviors, the motivations behind them and the actions they lead to, is at the heart of competitive intelligence.

Commerce and entrepreneurship cause individuals to form companies and cause those companies to compete with one another. In turn, nations and economies compete with each other. At the very root level of commerce, though, are individuals, and the choices they make. Understanding what other people are doing, how they think, and what their future plans are, are the key ingredients for successful competitive intelligence.

The most important attribute every good CI practitioner possesses is curiosity. This is probably why investigative journalists make excellent CI people. Without the relentless desire to understand why people do what they do and why things are the way they are, it's difficult to summon the necessary levels of insight to do top-grade CI.

Not everybody is extremely curious by nature, and your chosen occupation may have limited the extent to which you are curious on a daily basis. In fact, our busy lives often cause us to blot out any unnecessary stimuli so that we can concentrate. The good news is that it's fairly easy to develop or reawaken curiosity.

Give the following exercises a try.

EXERCISES: Improving Day-to-Day Curiosity

"I have no special talents. I am only passionately curious." Albert Einstein

1. This exercise is adapted from theater training. As part of the process of developing characters, acting students are often instructed to observe a person for some time. Then, they present to the class everything they know about the person they observed, including their speculation on the person's back-story or history. This is also a useful technique for developing curiosity for CI purposes.

   Go to a local coffee shop, mall or someplace you can sit comfortably. Take an hour to observe people, and ask yourself questions about them. For example: Where would you guess they buy their clothes? What might their occupation be? What are they choosing to purchase? Why might they be buying these particular items? What motivates them as consumers? Are they more oriented toward luxury or value? What do you suspect their attitudes would be toward various competing retail outlets or products?

2. This exercise builds on the previous one. Instead of focusing on people, you will consider companies. Think about a particular competitor or other company. Generate at least 25 questions about the company, more if possible. (Note that the purpose of this exercise lies in creating the questions; don't worry about the answers.)

For example: Why are they located where they are? How many retail outlets do they have? Who is next in succession for the top job? What is their organizational structure? What is the history behind their logo? Are their employees happy? What are their new product plans? Where did the CEO work previously?

Typically, CI questions revolve around companies and people. These exercises help develop or re-awaken your innate curiosity about the world around you. Remember, to gather good intelligence, you first have to ask good questions. Getting in the regular habit of practicing curiosity, and asking questions will better equip you as a CI analyst. Later in the book, you'll focus on getting the answers to your questions.

2

Sensory Awareness

"Use your five senses.... Learn to see, learn to hear, learn to feel, learn to smell, and know that by practice alone you can become expert." William Osler

Key Chapter Points:

- ❖ Intelligence is everywhere around us
- ❖ Effective CI involves using all your senses
- ❖ Methods to improve sensory awareness

The Role of Sensory Intelligence

In the previous chapter, you looked at why curiosity is an essential foundation skill for good CI. This chapter builds on that idea by introducing the role of sensory awareness.

Sensory awareness refers to the ways you can use of all your senses to gather intelligence. These include eyesight, hearing, touch, taste and smell, as well as intuition.

One of the pitfalls of CI is that it too often becomes a purely cerebral and analytical activity. This is limiting because it closes off huge amounts of intelligence you can gather using all your senses. If your goal is to get at least one percent ahead of your competition, then you need to be as ingenious as possible, using all the senses at your disposal.

We have evolved by using all our senses and abilities to anticipate and avoid danger, and to seek relative gain and advantage. You use your eyes, ears, touch, voice, taste, smell, experience and intuition to avoid and mitigate threats as well as to sense and exploit opportunities. CI practice is the same. Top-notch CI involves using all your senses and mental abilities to develop as full a picture of the competition and broader business environment as possible.

Intelligence Is Everywhere

> *"The universe is full of magical things, patiently waiting for your wits to grow sharper."* Eden Phillpots

An effective CI professional knows that intelligence is everywhere, in many forms. The following are all examples of intelligence:

- ❖ A snippet of conversation overheard in a line-up
- ❖ An observation of the facial expression of someone using a competitor's product
- ❖ The reaction you and your friends have to a competitor's advertising
- ❖ Knowing the hobbies of your counterpart in a rival organization
- ❖ A Google Earth satellite photo showing a half-empty competitor's parking lot
- ❖ A newspaper article about changing consumer trends

Because of its traditional link to market research, CI often falls into the trap of focusing too much on documented data and not enough on the more esoteric and powerful forms of intelligence mentioned above. The problem with this is that most rival firms will have access to comparable facts and figures.

The best CI analyst is one who can combine facts and figures with insights derived from sensory and intuitive sources to yield a unique view of the competitive landscape. The combination of different forms of intelligence can lead to a breakthrough in understanding and ultimately, a sustainable competitive advantage.

In a recent survey by the Society of Competitive Intelligence Professionals, "better integration of information from multiple sources" was most frequently cited as the change that would best improve CI activities in the coming year. (Source: SCIP report: "State of the Art: Competitive Intelligence" 2005-2006.)

Picking Up Signals: a hypothetical case

During a casual weekend visit to a store to check shelf placement for her firm's digital cameras, a CI analyst overhears a customer complaining that the newer cameras are no good for him because they don't have optical viewfinders. The young salesperson points out that the display on the unit they are looking at is 2.5", one of the largest available, with very clear image quality, so an optical viewfinder isn't necessary.

In exasperation, the customer explains that his middle-aged eyes can't focus close-up anymore. He tells the clerk he would have to hold the camera so far away to see the viewfinder that he couldn't compose the shot and that he's not about to fumble around for his reading glasses every time he wants to blast off a quick photo.

Armed with this piece of intelligence, the CI analyst goes to a nearby café with a wireless hotspot and gets on the Internet with her pocket PC. Going to Google Groups, she does several searches for mentions of farsightedness, difficulty viewing, optical viewfinders and the like. Finding nothing meaningful, she posts this question in the newsgroup *rec.photo.digital:* "Any older folks have trouble seeing to shoot a photo without an optical viewfinder?"

Returning home, the analyst gets back into Google Groups to find 12 responses, each echoing what the customer in the store had said. Sensing a breakthrough piece of intelligence, she emails her contacts in marketing and product development. In the coming days, the number of responses to the original newsgroup posting has swelled to several hundred. After confirming their hunches with focus groups and a quick web survey, the company is able to incorporate optical viewfinders – an old technology – into their next line of image-stabilized cameras, at minimal cost. They support this move with an ad campaign geared toward the older user – a lucrative target market. A chance piece of intelligence overheard in a store led to a breakthrough piece of strategy, but only because the CI analyst was there and knew the value of an innocuous remark.

Visual Observation

One of the richest sources of CI is your ability to observe people, places, things and events. For example, observing the behavior of people who consume your and your competitor's products is a useful and frequently overlooked source of intelligence.

Go to where your customers spend time interacting with your products, and see how they behave. In the example of a coffee shop, there is no limit to the number of observations you can make without attracting attention or even needing to interact with people. For example, take note of:

- ❖ What kinds of drinks do customers order?
- ❖ What size drinks?

- ❖ Do customers appear irritated, happy or indifferent in the lineup?
- ❖ How long is the waiting time on average?
- ❖ What do people put in their drinks?
- ❖ How long do people stay at the coffee shop?
- ❖ Who gets take-out?

By going to retail outlets, you can observe many things about your products and those of the competitor. Even a casual walk-through will reveal who has the best shelf space and whether or not competitors are offering any special discounts. You can check pricing, as well as how knowledgeable the staff is with the product features. You can also determine if the staff appears to have been given special inducements to push one company's products over another.

Touch, Taste and Smell

The senses of touch, taste and smell are very often overlooked in CI, which is unfortunate because these senses reveal a wealth of intelligence. Most of us at some time have probably bought a camera or similar gadget because it "felt right." You could try to get very analytical about what the notion of "feels right" means, try to boil it down into features such as weight, size, placement of buttons and so forth. Yet the greatest intelligence may lie in the intangible quality of feeling right.

If you conduct some focus groups, and consistently find your product feels better in the opinion of your customers, then you have a competitive advantage you can promote.

Similarly, if the focus groups show that the competition's product feels better, then you are at a competitive disadvantage. It is then that you may need to try to be more scientific about defining what the elements are that make the customer think the competitor's product feels better, so you can consider emulating or improving upon them.

Taste and smell can also be important sources of CI, especially in the consumer products arena, as the following exercise will show.

EXERCISES: Improving Sensory Awareness

1. Buy three or four brands of same size and flavor of yogurt (or similar product).

   ❖ Practice tactile CI by holding each container. Notice the shapes, the weight, how the lid feels. Which one feels superior to you? Why? Write down your observations, likes and dislikes.

   ❖ Now, look carefully at each one. Which one looks best to you? Why? Note your observations.

   ❖ Next, open each one. Was one easier or more pleasing to open than the others? Why? Note your observations.

   ❖ Continue this exercise with smell, appearance, taste, texture, and any other variables that occur to you.

   When you have finished, look through your observations and derive a list of attributes that a "winning" yogurt needs to have. You now have a working list of competitive success factors that you can confirm by vetting with focus groups or surveys. Even though your working list is subjective, you may well have the best intelligence in the industry at this point in time.

2. On a daily basis, practice noticing what you can see in reflected surfaces.

- ❖ Can you see the person walking behind you in store windows?
- ❖ Are they male or female?
- ❖ How tall?
- ❖ What are they wearing?

When someone enters a coffee shop where you are sitting, what can you learn without looking directly at the person, by looking at reflective surfaces like a cell phone screen, framed pictures, or windows?

These exercises help develop your visual abilities, as well the ability to gather intelligence without being obvious.

3

Numeracy

"Mathematics may be defined as the economy of counting. There is no problem in the whole of mathematics which cannot be solved by direct counting." Ernst Mach

Key Chapter Points:

- ❖ Getting comfortable with numbers
- ❖ Using rough estimates to gather intelligence
- ❖ Methods to improve numeracy

Getting Comfortable with Numbers

Another critical skill for CI is that of numeracy, or being comfortable with numbers. The UK Department of Education and Skills defines numeracy as follows:

Numeracy is a proficiency which is developed mainly in mathematics but also in other subjects. It is more than an ability to do basic arithmetic. It involves developing confidence and competence with numbers and measures. It requires understanding of the number system, a repertoire of mathematical techniques, and an inclination and ability to solve quantitative or spatial problems in a range of contexts. Numeracy also demands understanding of the ways you gather data, including counting and measuring, and presented in graphs, diagrams, charts and tables.

In short, numeracy involves being fairly comfortable with numbers, proportions, percentages, formulae and so on. Some may have been born with a knack for numbers and have nurtured these skills through education and career choices. The rest of us may have struggled to understand mathematics through school, some turning into genuine "math-phobes."

Like all skills, you can develop better numeracy. Fortunately, few CI practitioners need to be deep mathematical experts. Instead, what is required is an ability to think in straightforward mathematical terms. For example, here are some situations where numeracy helps gather intelligence:

- ❖ Being able to quickly eyeball a store shelf and estimate what proportion of shelf space is occupied by your products versus the competition's

- ❖ Being able to judge from graphs in annual reports whose sales are growing fastest

- ❖ Being able to determine the percentage represented by cost of goods sold in a column of financial statement numbers

- ❖ Being able to estimate the sales volume and revenue of a competitor from observed data

EXERCISES: Improving Numeracy

1. Think of a family of consumer products, like laundry soap, or canned soup. Practice going into retail stores, and without spending more than a few seconds, try to estimate the approximate share of shelf space held by each major brand, in percentage terms. Keep practicing this in different stores, so you develop a facility for quickly sizing up situations in mathematical terms. Occasionally, when time permits, do a quick count of the facing boxes or cans and check your percentages. The more you practice, the more accurate you will become.

2. Place yourself in a position to observe a small retail location, for example a coffee shop or similar business that takes in cash and has steady customer turnover.

   - How many customers come and go in an hour?
   - Can you guess by observation at the average purchase size? What would the hourly revenue be approximately?
   - How many employees are there?
   - What would you guess they are paid by the hour?

   From this information, develop a rough picture of whether or not the company is taking in more money than it is paying its staff. What you have developed is a model of the business, which can now be refined with such factors as an estimate of rent, more observation periods to identify peak traffic flows, and corresponding changes in staffing levels and so forth.

   The more you practice building models, the easier it gets. There is little qualitative difference in modeling a small retail store and a large corporation. The more you are able to understand about your competitor's business model, cost structure, revenue flows and other details, the better you will be able to identify and exploit any weaknesses inherent in their business.

4

Research Ability

"Discovery consists in seeing what everyone else has seen and thinking what no one else has thought." Albert Szent-Gyorgyi

Key Chapter Points:

- ❖ Quickly and efficiently accessing existing information is a key skill
- ❖ Most of the information you need for CI already exists
- ❖ Ways to improve your research abilities

Gathering Information

Another core skill in CI is the ability to quickly access and use sources of existing information. A good CI practitioner will be able to think of a dozen sources of information about a particular problem. Sources of information can be as varied as the following:

- ❖ Online databases
- ❖ Company annual reports and other filings
- ❖ Published articles and newspaper stories
- ❖ Business directories
- ❖ Statistics
- ❖ Market research reports

Most of the information you need for effective CI already exists in some form. The key is to be able to access the information fast, by keeping warm in your mind the most helpful sources and starting points.

One of the best sources for information on a publicly-traded company is the company itself. For example, a company website will usually have a section detailing the background, mission and key players in the company. As well, most companies have an investor relations section on their website offering a convenient short cut to press releases, regulatory filings, and annual report information. Chapter 7 will explore Internet-based sources of information in more detail.

Other excellent sources of information include government reports, bank and investment publications, trade magazines, online proprietary databases, annual company reports. You can find most of these resources at any good college or public library. For now, the key is to develop as broad a sense as possible of the information sources that are available. The following exercises will acquaint you with a wide array of information sources.

EXERCISES: Improving Research Abilities

1. Go to a local public or college library, and find the business librarian. Ask for a guide to using the library for business research. Time permitting, many librarians will be able to give you an overview of the resources available, both printed and online. At a minimum, there will be a printed outline of the resources available.

   Spend an hour or so trying to gain as complete a picture as possible about a particular company, including its structure, performance, key markets, industry trends and so forth. Ask the business librarian for assistance. Make a note of the most useful information sources you found, for future reference.

2. Choose a large, publicly-traded company to investigate. Visit their website and mine the site for all the information you can unearth about the company's leadership, strategic plans and recent news. Tap into items like "careers" to see if the types of people being recruited can offer insights into where the company is headed.

5 Ethics

"Worry more about your character than your reputation. Character is what you are, reputation merely what others think you are." John Wooden

Key Chapter Points:

- ❖ Maintaining the highest ethical standards leads to the best intelligence
- ❖ The Society of Competitive Intelligence Professionals Code of Ethics
- ❖ Ethical dilemmas and blunders
- ❖ Exercises to develop applied ethics

An Ethical Way of Being Competitive

Nothing about top-notch CI has to be unethical. You can actually gain superior intelligence by following

31

the strictest ethical protocols. The challenge is that you need to use your mind in a more effective and disciplined manner, especially the parts of your mind that you might not associate with intelligence, like intuition. Lying, deception, stealing, bending or breaking laws are not only unnecessary, they are symptoms of mental laziness.

An Ethical framework

CI is often equated with corporate espionage, and with breaking the law. Regrettably, the profession itself may be somewhat responsible for this. In order to make the subject more appealing to business students and other audiences, it is sometimes tempting to play up a little of the covert, "racy" image of CI. In truth though, probably few other professions govern themselves with such a stringent code of ethics as the CI Profession. One important industry association is SCIP, the Society of Competitive Intelligence Professionals, which maintains the SCIP Code of Ethics, presented below.

SCIP Code of Ethics for CI Professionals

- ❖ To continually strive to increase the recognition and respect of the profession
- ❖ To comply with all applicable laws, domestic and international
- ❖ To accurately disclose all relevant information, including one's identity and organization, prior to all interviews

- ❖ To fully respect all requests for confidentiality of information

- ❖ To avoid conflicts of interest in fulfilling one's duties

- ❖ To provide honest and realistic recommendations and conclusions in the execution of one's duties

- ❖ To promote this code of ethics within one's company, with third-party contractors and within the entire profession

- ❖ To faithfully adhere to and abide by one's company policies, objectives, and guidelines

Source: Society of Competitive Intelligence Professionals (scip.org)

Some Ethical Blunders

Recent history has shown us a number of ethical blunders. In one case, a company was ordered to pay a settlement to its rival, after it was revealed that the company had used the knowledge of a rival's former employee to gain access to the competitor's computer systems. This unauthorized access was used to uncover confidential information that would have unfairly benefited the company.

In another case, a manufacturer was trying to uncover the source of leaks to the media from within its board ranks. Using investigators, the company conducted background checks on several employees, their spouses and children.

As part of the investigation, they obtained phone records under false pretences, a practice known as pretexting. The case has to-date resulted in several high-profile resignations and the multimillion dollar settlement of a civil lawsuit.

Both these cases are obvious breaches of ethical business standards. While it may be tempting to go after information because it's available, it seems very short-sighted to risk massive damage to company reputations by taking the easy way out.

It is not necessary to engage in unethical behavior to obtain effective competitive intelligence. None of the tools presented in this book are illegal or unethical in themselves. Needless to say, some techniques can be used unethically or illegally, but this is not necessary. Extremely useful intelligence is around us everywhere; you can obtain it legally and ethically to give your firm a sustainable strategic advantage.

EXERCISES: Developing Applied Ethics

1. Imagine you and your competitors are both exhibiting at a tradeshow and you would like to find out about any new product launches that are on the horizon. You will be hiring marketing students from a local college to gather intelligence. Develop a staff-briefing document for a trade show intelligence-gathering operation. How will you equip, script and brief these students so that they can gather the intelligence you require, while still fulfilling all the requirements of the SCIP Code of Ethics?

2. Consider a potential CI situation in which you might find yourself. For example, there is a piece of intelligence you would really like to obtain. How far would you go to obtain it? Some possible approaches are sketched out below. Where you end up on the spectrum below represents your personal code of ethics. Does it comply with the SCIP Code of Ethics?

   ❖ Breaking the law to get the intelligence

   ❖ Paying someone to get the intelligence knowing they would likely break the law

   ❖ Hinting to a subordinate that obtaining the intelligence would be a career-enhancing move

   ❖ Estimating the intelligence only from legally-available sources

   ❖ Giving up on the intelligence if you could not obtain it legally and ethically

PART TWO:
Applied CI Skills

Applying What You've Learned

Having honed the core skills for CI in the previous chapters, you are now ready to explore some applied tactics and techniques of CI. It is still important to keep practicing the core skills developed in Chapters 1-5; even experienced CI practitioners must do this in order to keep sharp.

The techniques in the following chapters have been chosen because they are fast, cost-effective and can be done ethically, by a single person if necessary. In fact, you can gather a vast amount of information right from your desk.

6

Industry Analysis

Key Chapter Points:

- ❖ Industry analysis as the starting point for CI
- ❖ Porter's 5 Forces model
- ❖ Identifying drivers of competitive success
- ❖ Threat levels and counterattack strategies
- ❖ CI reporting matrix
- ❖ Exercises to practice industry analysis

Finding a Place to Start

All CI needs a starting point, and it is here that many efforts fall short. The bias toward so-called facts and hard data leads many CI efforts to stall before starting. In Chapter 2, you developed a working list of competitive success factors in the flavored yogurt industry. You gathered data using activities involving touch, smell and taste and made other observations that were subjective and personal in nature. Nonetheless, this is valuable intelligence, and a very worthwhile starting point.

How likely is it that your own observations are going to differ very much from the typical consumer? You can always "harden" your observations through more testing, such as focus groups followed by a quantitative survey. But the key is that you have a starting point, which may be more than your competitor has, so you already enjoy an advantage. This chapter will look at starting points, because these form the basis for all successful CI.

Things to Look For

There is no limit to what forms of intelligence are useful as starting points in CI. In very well-studied industries like retail, there are countless measures, including AC Neilson reports that can give a ready-made set of benchmarks. But for many industries, you need to create your own set of competitive indicators. For this, you need some starting point. As you saw in the yogurt exercise, any starting point can form the basis for a comprehensive CI effort – the key is to start.

Industry Analysis: the starting point

One useful place to begin is with the industry overall. What drives competitive success in your industry or local market? To be effective in analyzing this, you need to dig well below the surface. Your first instinct may be to rely on the "usual suspects" like price, sales volume, market share, profitability and so on. However, to develop an effective base for CI, you need to go deeper and try to get at the root drivers of competitive success. What you think of as drivers of competitive success may in fact only be symptoms of deeper root issues.

For example, it may appear as though the lowest price wins in your industry. However, if you dig deeper, you may find that companies are constrained in their pricing by their cost structure. Looking further, you see the biggest difference between the cost structures of the leading and lagging firms is production labor.

You need to analyze this further. Is there a difference in headcount per unit produced? In wage rates? In job structure? Are there union and non-union shops? After drilling down deeply enough into the problem, you may find that the leading firms have a particular piece of machinery that allows for greater productivity. It is this, then – the presence and quantity of this piece of equipment – that should be seen as the driver of competitive success, not price as the surface analysis first suggested.

You may also find through human intelligence (perhaps one of your workers used to work for the competition and volunteers the insight that there was an incentive given in time-off for greater productivity) that incentives play a key role in lower relative labor costs. So it is the presence and nature of these incentive programs that should be treated as a driver of competitive success.

Porter's 5 Forces Model

A useful tool for determining the key drivers of competitive success is to use the 5 Forces model popularized by Michael Porter in his book *Competitive Strategy: Techniques for Analyzing Industries and Competitor* (Free Press, 1998). Porter's model looks at the forces in an industry that shape competition. These are:

- ❖ Threat of new entrants
- ❖ Bargaining power of buyers
- ❖ Bargaining power of suppliers
- ❖ Threat of substitute products or services
- ❖ Rivalry among existing firms

A surface analysis using Porter's 5 Forces will yield insights such as: "The power of buyers is strong because there are many competitive products and switching costs are minimal." This is where typical business school courses will leave off. To effectively use the 5 Forces for CI purposes though, you need to dig deeper, and ask yourself questions such as:

- ❖ Given what you know from the 5 Forces, what can you do to increase the switching costs for your customers?
- ❖ How can you do this at least slightly more effectively than your competitor?

Once you've identified the key drivers of competitive success, focus your efforts on the ones you have the greatest ability to influence. The best drivers to focus on for CI purposes are:

- **Changeable:** You have control over the driver. If you and your competitors all face the same external force and nobody is able to influence it, then it isn't worth focusing on for CI purposes, no matter how important it is to your business.

- **Measurable:** You can develop a reasonable measure of the driver.

- **Differentiable:** You are in a position to gain relative advantage over your competitors on this driver.

After you have isolated the key drivers of competitive success, you need to look at your own firm, at your internal competencies. You are again looking for areas where you have, or can gain, a relative advantage over the competition. If success depends on having a particular technical expert, is yours a bit better than the competition? If not, can you improve the skill set of your expert with training, or by recruiting another person?

Next, you need to consider what external forces are in play in your business environment. These can range from government regulation to changes in consumer preferences. These are the factors over which you typically have little or no direct control. However, you may be able to gain relative advantage over your competitors by understanding these forces a little better, or by maximizing your mobility a bit better than the competition. For example, if legislation affecting your industry might change, can you get yourselves in a position to find out about the change before the competition?

Similarly, can you predict changes in consumer behavior faster and better than the competition by having a consumer panel you regularly poll for changing attitudes and preferences?

Once you have identified the key drivers of competitive success, looked at how you rate on these versus your competition, examined the external forces in play in the business environment, and made a candid assessment of your internal strengths and weaknesses, you are ready to pull your intelligence together and consider threat levels and counterattack strategies. Once you have a complete picture of your competitive landscape, you can then use a CI matrix, shown in the exercise below, to summarize your findings.

Threat Levels

For each major competitor that you have analyzed, you need to come up with an assessment of how threatening they are as a competitor. A simple way of doing this is to assign each major competitor a threat level score from 1 to 10, where 1 means "not a threat" and 10 means "a great threat." It is important to avoid the temptation to get too analytical or precise in assigning the 1 to 10 threat level to each competitor. Instead, treat it as a rough guide to identify who you should be worrying about the most.

Quickly examine the key drivers of competitive success and see whether the competitor is winning on most of them, and if so, by how much. If your company is being soundly beat on most of the drivers, then that competitor would be an 8, 9 or 10. If your company is winning on most of the drivers, then the competitors would be a 5 or less.

Do this for each competitor, and then rank the competitors from most threatening to least threatening. It makes sense to focus your efforts only on those competitors that pose a substantial threat, say those with threat levels of 6 or higher. The lesser threats can be dealt with on a less urgent basis for now.

Counterattack Strategies

Now that you have examined each competitor in depth, and determined which ones are the greatest threats, you can give attention to counterattack strategies. These strategies and tactics are the things you will do to accomplish one of the following:

- ❖ Enhance your competitive advantage in the areas in which you lead
- ❖ Reduce your weakness in the areas where others are leading

When considering counterattack strategies, it is important to keep these brief, tactically-focused and realistic. If your disadvantage is price, simply lowering your price is not necessarily a viable option, given your cost structures and margin requirements. A more realistic strategy may be to enhance perceptions of being the higher-value product, through your marketing communications. Another viable counterattack may be to offer a price break to your top-tier clients in order to build loyalty. The more specific, realistic and actionable your counterattack strategies are, the more success you will have communicating and implementing them.

Tying It All Together: the CI matrix

The exercises that follow will lead you through the process of crystallizing this chapter's activities into a single sheet of paper, the CI matrix. The matrix is an extremely powerful tool for organizing and presenting a company's competitive landscape in a single snapshot.

The matrix focuses attention on a company's main competitors, assessing each one on the major drivers of competitive success, and details the threat level of each competitor. Finally, the matrix outlines the counterattack strategies that relate to each competitor. The beauty of the matrix is that it is concise and clear, and can easily be distributed to all relevant personnel. The matrix shows at a glance where your energies need to be focused, and what you must do to maintain or gain a competitive advantage.

By working through the following three exercises, you will learn how to build this powerful reporting tool.

EXERCISES: Practicing Industry Analysis: building a CI matrix

1. For your industry, brainstorm the forces of competition using the 5 Forces model. Develop a list of what you think are the key drivers of competitive success, keeping in mind the desired qualities of being changeable, measurable and differentiable. Isolate 4-5 key drivers that you think are the most meaningful. Use these to begin building a matrix in the following form:

	Driver 1	Driver 2	Driver 3	Driver 4	Threat Level	Counterattack strategy
Your Company						
Competitor 1						
Competitor 2						

2. Once you have developed your key drivers, decide which competitors to focus on. You may have many competitors, but chances are there are only a few that really give you a headache and keep you awake at night. These are the ones to include on the matrix. Next, find or develop data for the key drivers you have chosen. This is more challenging than it looks, since you may need several sources and calculations to arrive at an estimate of the driver in your competitors.

3. Next, assess the threat level of each major competitor from 1 to 10. Consider dropping the least threatening competitors from the matrix at this point. For the most threatening competitors in the CI matrix, develop counterattack strategies directed against each competitor, remembering to keep the counterattack strategies focused, realistic and achievable.

7

Internet Competitor Analysis

Key Chapter Points:

- ❖ The Internet as a CI tool
- ❖ Using the Internet more strategically
- ❖ Accessing the hidden Internet and Web 2.0
- ❖ Exercises to practice Internet CI techniques

The Web of Intelligence

The Internet is now a broadly understood resource for finding information and this book will not address the basics of doing searches, which are well-covered elsewhere. Nor will it offer an in-depth guide to doing online research. Instead, we offer a window into the huge variety of CI resources available online as well as a look at some of the newer uses of the Internet for CI.

The tools in this chapter are those that can give you an edge over your rivals, helping you gain better information, faster. These include Web 2.0 applications such as blogs, social networks, social bookmarking and tags, virtual worlds as well as older tools such as meta-search engines, aggregators and bots.

Google for CI

Google is likely the best basic tool to use when searching for CI. In addition to the usual search tools, Google offers several specialized ways to search that can prove useful for collecting CI. For example, Google Groups offers a searchable archive of more than 700 million Usenet postings from a period of more than 20 years. There are a few ways a CI practitioner can use Google Groups (**groups.google.com**). First you can search the archive for information on your topic by doing an online search. Another alternative is to sign up to relevant groups and see what people say about your competition or your industry. You can also "seed" a discussion by posing a question and seeing what responses you get. By posting a comment along the lines of "I hear Company XYZ is pulling its line of digital cameras," you may be able to ferret out specific pieces of intelligence, which can help you gather information on evolving competitor activities.

Google Labs showcases BETA versions of new Google products such as Google Trends and Google Experimental Search.

Google Earth offers maps and 3-D images for complex or pinpointed regional searches.

Google Scholar provides a search of scholarly literature across many disciplines and sources, including theses, books, abstracts and articles. These can be useful in helping you collect well-researched data. It is useful to conduct a "search by date" in order to find the most recent articles.

Google Patents allows a search of over seven million patents. Another useful site is the United States Patent and Trademark Office (**uspto.gov**) which has a searchable database of pending, registered and dead American patents and trademarks.

Google frequently adds new tools and products that may help you collect intelligence. Keep up to date by checking google.com/options.

Getting Beyond Google

Once you have exhausted what Google has to offer, explore other tools such as meta-search engines and services that crawl through the "Deep Web." There are many different kinds of search engines available; for simplicity's sake, we will narrow our focus to three free services.

Meta-search engines employ several search engines simultaneously, increasing the chances of finding information missed by a simple Google or Yahoo search. Some free services to try are:

- ❖ **clusty.com:** A meta-search engine that queries top search engines, combines the results, and generates an ordered list based on comparative ranking.

- ❖ **surfwax.com:** A meta-search engine that allows you to select how many results you want, ranging from 25 to 500-plus. The results also come with Site-Snaps (TM) offering real-time summaries of each site found.

- ❖ **sputtr.com:** In addition to searching engines like Yahoo and Ask.com, sputtr queries wikis, dictionaries and encyclopedias as well as social websites like youtube.com and shopping sites like amazon.com.

Be aware: Search technology changes all the time. To find a list of other search engines available, visit **searchenginewatch.com.**

The Invisible Web

The Deep or Invisible web is huge. There are tens of thousands of pages of online data buried in databases and websites that search engines often miss. One way to access these pages is to include the term "database" with your other search words. Another is to search using Google Scholar or the Librarians' Internet Index (**lii.org**). An additional service offering dynamic database searches is Complete Planet (**aip.completeplanet.com**).

A complete explanation of how to access Deep Web resources is beyond the scope of this book; you can find a more comprehensive explanation at The Online Education Database (**oedb.org/library/college-basics/research-beyond-google**).

RSS and Web Aggregators

Several free aggregator services make it easy for you keep track of people, ideas and trends. For example, "Really Simple Syndication" or RSS feeds collect updates from blogs, podcasts, newspapers and magazines. The updated information is sent your email inbox or to a reader you install on your desktop. There are many readers and formats available, including: **feedburner.com, reader.google.com** and **newsgator.com**.

After you've set up your RSS aggregator, it is easy to subscribe to the publications you want to track. For example, say you are interested in tracking people and trends related to organic food. Once you have located online publications, blogs and podcasts containing content on organic food, subscribe to their RSS feeds and get updates sent to your reader.

Some sites with useful RSS feeds for business include:

- Bizjournals.com
- Financial Times (ft.com/servicestools/news-tracking/rss)
- Forbes.com
- MarketWatch.com
- Wall Street Journal (online.wsj.com)

Searching News Sources

There are two ways to collect CI from news sources: By a direct search or by setting up news alerts to be delivered to your email inbox or RSS feed reader. Google News (**news.**

google.com) and Google News Alert offer both services. Simply enter your search terms for current news and sign up for news alerts by clicking on the "News Alert" link at the bottom of the search page. Other places to search for news are: reuters.com, newsiness.com and News Search Portal (**nieuwsbronnen.com/newssearchportal**).

Note that it may take several attempts for you to narrow down an effective set of search terms. If you don't get good results with your first terms, find alternatives. Practice makes perfect!

There are also many sites that search television and radio news transcripts and offer video and audio feeds. Four free services are: **youtube.com, video.google.com, metacafe.com** and **video.yahoo.com.**

Additionally, there are several fee-based services including: **TVeyes.com, criticalmention.com** and **tvnews.vanderbilt.edu.**

Searching Blogs

There are over 100 million blogs in existence and many of them contain information that can help your business. Not only can you track what is said about your company's image and products, but you can also collect information on competitors and trends via the RSS feeds found on most blogs. (There are few richer sources of information than the blog of a disgruntled ex-employee or an angry customer.)

Again, there are many blog searchers available; the following three are good places to start. Try **blogsearch.**

google.com which works just like Google search. Simply type in your key words and hit the search button. Similar blog search engines are found at **Blogdigger.com** and **IceRocket.com.**

Technorati.com is another blog search engine. By the end of 2007, it had indexed over 111 million weblogs and 250 million pieces of tagged social media. There are many different ways to search Technorati; the most basic is to search by keywords and phrases. Another method is to search by "tags". For example, entering the key words "organic food" into a tag search returns not only all blogs that have content on organic food but also gives a list of related tags such as "green", "sustainable" and "agriculture". (Note that tags also offer the opportunity to see what people are blogging about the most at any given time.)

A third way to use Technorati is to set up a Watchlist. This customized reporting tool delivers updates on your keywords several times an hour.

Blogpulse.com, a blog tracking service offered by Nielsen Buzzmetrics, is of particular interest to those gathering CI. In addition to searching blogs, it offers trend analysis and conversation tracking features. For example, by entering the key words "organic food", you can see what percentage of blog posts were about this topic for each month of the past year, and can follow the blog conversations held on this topic. Blogpulse also has a "featured trends" page that focuses on specific topics in science, health, news, sports and so on.

Searching For and Tracking People

Forbes People Tracker (**forbes.com/cms/template/peopletracker/index.jhtml**) allows you to track over 120,000 executives as well as members of the Forbes "rich and celebrity" lists. You can also set up email alerts on individuals and companies.

Pipl.com searches the Deep Web for personal profiles, public records and other people-related documents are stored in databases. **ZoomInfo.com** offers free and fee-based searches on over 37 million people and three million companies. **Wink.com** searches social networking sites such as MySpace, Bebo, Friendster, LinkedIn and Live Spaces to help you find people by name, location, school, work, and interests.

Searching For Experts

Several online sites will help you locate experts in your area of interest. They include **Experts.com, Profnet.com** and **Authoratory.com.**

Using Shopping Bots to Collect Data and Opinions

Shopping Bots are a useful method to track competitor products, pricing and catalogues. For example, **mySimon.com, Shopzilla.com** and **Shopping.com** search out price, shipping and warrantee information from thousands of merchants. Other services such as **Epinions.com,**

RateItAll.com and **cnet.com** also offer customer ratings so you can see how your competition stacks up.

Keeping Track of Competitors' Websites

An easy way to monitor the changes your competitors make to their websites is to sign up for either a free or by-subscription service that tracks content updates. Examples of these services are:

WatchThatPage.com is a free service that lets you collect updates from selected websites. When new content is added, you'll receive an email.

ChangeDetect.com monitors changes of web pages and emails color-coded highlights of what changed.

EyeOnWeb.com is a fee-based service similar to the two services mentioned above.

Copernic Agent (**copernic.com/en/products/agent/index.html**) is a by-subscription customized monitoring service that offers tracking of both visible and hidden web pages.

Social Networking and CI

Social networking is a burgeoning phenomenon. People are flocking to sites such as MySpace, Facebook and LinkedIn by the millions and posting vast amounts of information about themselves on a daily basis. As a CI analyst, you can learn a terrific amount about your competitors by monitoring these sites, gaining insights into your competitors' education, hobbies, social activities and so forth. These pieces of information can provide valuable clues to personality, philosophy and future plans.

In addition, if you are able to connect to these people, you can also see who their connections are. It can be very revealing to see who is connected to whom and may tip you off to activities such as pending mergers, job changes and other interesting tidbits of competitive information.

This aspect of using social networking sites to profile individuals is in its infancy, but it can still be a powerful tool for you to deepen your understanding of the strengths and weaknesses and potential vulnerabilities of the people you are competing with.

EXERCISES: Practicing Internet CI Techniques

1. This exercise allows you to mine the many discussion groups on the Internet for CI. Go to Google, and click on the Groups option. This permits you to search a myriad of discussion groups that exist on the Internet, but which can't usually be found through a Web search.

   ❖ Type in your company name, or your product name, and hit search. What are people saying about your company, and its products?

   ❖ Now key in your competitor's name, and analyze the comments. You have just discovered a free resource that can alert you to problem areas in your own company's perceived image, as well as vulnerabilities in your competitors. Use this tool regularly to scan for intelligence.

2. See chapter 13 "Executive Profiling" for more ideas of how you can use online social networks to create a profile of your competitors.

8

Qualitative Market Research

Key Chapter Points:

- ❖ Primary market research as a CI tool
- ❖ Introduction to qualitative research
- ❖ Moderating and analyzing focus groups
- ❖ Using focus groups and in-depth interviews for CI
- ❖ Ways to develop qualitative research skills

Subjective Intelligence

Traditional forms of market research such as surveys and focus groups are very useful in competitive intelligence. These tools allow you to go to the market and learn how both you and your competition are perceived. You can also learn what people like and dislike about your products and those of your rivals. However, using a professional firm to conduct market research can be expensive, so in this chapter you will consider ways you can use these tools inexpensively, either by doing the research yourself or by using cost-effective alternatives such as college market research students.

Primary Market Research Methods for CI

There are two main approaches to primary market research: Qualitative (focus groups and in-depth interviews) and quantitative (surveys). You can use both of these approaches for CI purposes; they needn't be expensive or difficult. You can find in-depth information on market research methods at sites such as **MarketResearchWorld.net**. A simplified version follows.

Qualitative methods like focus groups are used to explore an issue in depth while speaking with relatively few people, with results that are not statistically valid. In other words, you can get many insights from qualitative research, but you can't then turn around and project the results to a larger population, and infer that everybody feels the same way.

The goal of quantitative research, by contrast, is to make statistical projections from a sample of people. This is done by choosing that sample carefully enough that it mirrors a larger population, and then asking the people in the sample a uniform set of questions. In quantitative research, you explore issues in relatively little depth, with a large group of people, so that you can then say with a known level of accuracy that an entire population feels similarly to the people in your sample.

This is a gross simplification of a very scientific field, but it is the essence of primary market research. Remember, if you want *scientifically* valid research, hire a professional. That said, let's examine how you can use these tools quickly and inexpensively to gather competitive intelligence in ways that may not be entirely scientific, but can still yield valuable intelligence. The caveat is that you should not rely on information generated from "fast and easy" approaches to research for important decisions without first confirming your findings with professionally-conducted research.

Qualitative Research

Focus groups are a fast way to learn a lot from a small group of people. A focus group is a discussion led by a moderator, whose job is to guide the discussion along a path that results in the group exploring the issues that are important to you. Be aware: moderating a focus group is a challenging exercise to do well.

The process of setting up a focus group has several phases:

- ❖ Determining objectives: What do you want to learn from the group?

- ❖ Developing the discussion guide: What topics will you cover? In what order? How much time will you allow for each topic?

- ❖ Recruiting: Who should be in your group? How will you find them? How will you attract them to the group (incentives)?

- ❖ Conducting the group: How will you keep the discussion on track? How will you make sure everyone speaks? What logistical considerations must you take care of (room booking, taping, food, distributing incentives)?

- ❖ Analyzing and presenting the results: What major themes emerged during the discussion? What should you do based on what you've learned?

In an ideal world, you would conduct a number of focus groups that covered a cross-section of your customers and potential customers. You would always do more than one group in order to provide a measure of scientific control. You would also use the services of a professionally-trained moderator. Ideally, none of the participants would be known to the moderator, or to each other, to minimize bias.

In the rough and tumble world of competitive intelligence, a single focus group comprised of people you know, that you moderate yourself may be all you need to gain insights that put you slightly ahead of your competition and give you a sustainable advantage. Here then, are some guidelines— and pitfalls to avoid— in doing your own focus group research.

Determining Objectives

A typical focus group runs around ninety minutes to two hours, with approximately six to twelve people in attendance. It is important not to overload the group with too many discussion topics. In a couple of hours, you can only hear from each participant for around ten minutes, which is not much. It is better for each person to be able to expound on fewer topics for a greater amount of time.

A good guideline is to subtract ten minutes for the introduction and five minutes for the wrap-up from the total time available from the group, then split the remaining time between around four to six topics, weighting the time according to the importance and complexity of each topic.

Developing the Discussion Guide

The discussion guide is a one to two page "road map" for the focus group. Its purpose is to remind the moderator of the topics to be covered as well as the time allotted for each topic. The guide is not shared with the focus group participants. It should be skeletal in nature, and laid out in a large font with plenty of white space so the moderator can easily glance down to see what the next topic is, as well as the time available for each topic. Any special instructions relating to the introduction and wrap-up of the session should also be included. If there are any special techniques to be employed, such as the use of photographs to kick off a discussion, these instructions should be noted as well.

Sometimes, the discussion guide only represents your "best guess" as to how the group will proceed. An excellent moderator is able to be flexible and adapt the discussion guide on the fly in response to an unexpected but useful turn of events, and still cover all the topics on time. (For more information on how to write a focus group guide, see *Focus*

Groups: a practical guide for applied research by Richard Krueger and Mary Ann Casey, Sage Publications, 2000.)

Recruiting

Who should you include in your focus group? The answer depends on your objectives. If you want to determine what customers think about your products and those of your competitors, then you want to attract a blend of people who use your products and people who use each of the competitors' products. Also, since people of different ages, education levels, genders and so forth have differing opinions, you'll want a cross-section of different kinds of people. If you are trying to get it all done with only one or two groups, the challenge of group composition can be daunting.

How many people should you recruit? Opinions on this vary among professionals, but the typical range is from six to twelve people per group. The argument for fewer people is that you'll learn more from each person and can get into more depth than with a larger group. The argument for more people is that you can get a greater variety and range of opinions expressed. Smaller groups may suffer from a lack of energy and group dynamics, while a large group can suffer from silent free-riders, as well as control problems for the moderator. Personally, I favour smaller groups, from six to eight people.

Another thing to factor in is no-shows, which may happen in 25% or more cases. A good rule of thumb is to recruit ten people, hoping for eight to show, which allows as many as four people to not show up and still leaves you with a workable group of six people.

How should you get people to attend your group? Incentives are the norm in focus group research, with cash being the favorite. At the time of publication, $60-75 per person for between 90 and 120 minutes of time is probably fair and should be effective. You may get people to come for as little as $40-50, but it is generally better to spend the extra money to give a meaningful incentive. The higher rate usually buys you a greater degree of seriousness and professionalism. Current prevailing rates can be determined by querying local market research practitioners or by searching the Internet. For example, **craigslist.org** often advertises upcoming focus groups.

When budgets are too constrained to allow for cash incentives, you may be able to offer non-cash compensation in the form of products, discounts, or entry in a prize draw. If students are a viable source of opinions for you, you may be able to buy their participation with free food and refreshments, as well as offering a learning experience. The bare minimum in incentives should be plenty of food and refreshment, as well as the promise of an enjoyable discussion.

Conducting the Group

Where should you hold your focus group? The main things to consider are the facility, whether to use video or audio taping, and what types of food and refreshments to provide.

For most groups, a boardroom table seating arrangement is best, since discussion is enhanced when everyone can see and speak to each other. If a boardroom table is not available, you can rig up something similar by making a rectangle of smaller tables and throwing some tablecloths on them.

If your budget permits, a market research house can provide a room that is specifically designed for focus groups, and can supply videotaping as well as food and refreshments. However, you can conduct good focus groups for less money using improvised means. Hotel meeting rooms can be an affordable alternative, as can community centre rooms, library meeting rooms and the like.

If possible it is best not to hold the group on your own company premises, since people may tend to "accentuate the positive" once they realize who is paying them, hoping to tell you what you want to hear. The more disguise you can bring to the process the better, in terms of getting balanced opinions.

Leading the Focus Group

To conduct a successful focus group, the moderator needs many key skills. He or she must be able to:

- ❖ Build rapport with people: Be approachable, friendly, dress similarly to the participants, use their language.

- ❖ Stimulate discussion: Introduce each topic without dominating, get everyone to participate through verbal and non-verbal encouragement and active listening.

- ❖ Control group dynamics: Know when to back off and allow discussion and when to intervene and control dominating or disruptive personalities.

- ❖ Keep the discussion on track: Keep to the timelines for each topic, while remaining flexible enough to sense when a fruitful discussion should be allowed to continue overtime if necessary.

- ❖ Remain impartial and objective: Leave your own biases outside the room, never inject your own opinions, either verbally or non-verbally.

- ❖ Read verbal and non-verbal signals: Understand when a person's tone-of-voice and body posture are indicating that they have more to say, are upset, or are not being truthful. Know when to push and probe further and when to back off.

- ❖ Remain flexible and adaptive: Be able to go with the flow when the discussion goes off track in a surprising yet very productive way. Be able also to get back on track and cover all the required topics on time.

If you don't feel comfortable conducting the group, try to find an inexpensive alternative. For example, many college market research courses require students to perform research as part of their course fulfillment. A call to a local market research instructor will usually bring an enthusiastic response and a volunteer student moderator.

At the end of the group, thank everybody and distribute the incentives. It is also a good idea to see if anyone knows someone (themselves included) who would be willing to be involved in future research, since this makes recruiting easier the next time.

You should make a video or audio tape of each focus group session. This allows you to do more in-depth analysis later. A moderator can't possibly cope with the conduct of the group, and still have enough attention left over to note everybody's comments, tone of voice and body language. By making a taped record, you can analyze the group at your leisure and draw deeper understanding from the discussion.

Analyzing the Results

How do you analyze the results of the focus group? Analysis of qualitative research is more art than science. If you have personally moderated the group, all you may need to do is go home and reflect on what you heard, revisiting the recording as necessary to confirm your memories and get verbatim quotes if desired.

However, if you've done several groups, or the information is more complex, then you need to take a few extra steps. A simple method of analysis is to note the different kinds of opinions expressed in response to each topic and group them into "clusters" or "families" of similar opinions. Then you can count how often an opinion was expressed to get a sense of what was most important to people. Software is available to assist with this, but here is a cheap and easy way to do qualitative analysis:

- ❖ Replay the tape of the group
- ❖ For each topic, jot down each expressed opinion separately on a sticky note
- ❖ Stick the notes on a wall or whiteboard
- ❖ As similarities are noted, group the yellow sticky notes together into clusters
- ❖ Consider each cluster of notes
- ❖ What is the main theme of each cluster? These themes are the basic findings of your focus group
- ❖ Develop conclusions and recommendations based on these findings

Often, just the impressions you have formed through moderating the focus group will be enough intelligence for your particular purposes. In that case, you may not need to perform any further analysis until time permits. Remember, fast intelligence often beats in-depth intelligence, especially when you are trying to gain an edge on your rivals.

In-depth Interviews

An in-depth interview is quite similar to a focus group, with the key distinction being that the moderator only interviews one person at a time, rather than a group of people. This approach can make sense for situations where:

- ❖ The topic is such that people wouldn't want to talk in a group setting, due perhaps to embarrassment, privacy or competitive concerns
- ❖ The topic requires deeper discussion and interaction with the respondent than would be permitted in a focus group
- ❖ The topic is very technical in nature
- ❖ The people required to interview are rare, time-pressured or expensive, making the scheduling of a group session unrealistic

The logistics of conducting in-depth interviews are basically the same as for a focus group. Setting objectives, designing an effective discussion guide, arranging an interview location, managing the discussion and compensating the participants are all as important in in-depth interviews as they are in a focus group.

Other Uses of Qualitative Research

You can also tailor qualitative research to specific CI situations, including:

- ❖ Taste tests
- ❖ Customer needs analysis
- ❖ Product comparisons
- ❖ Prototype testing
- ❖ Idea generation
- ❖ Concept testing
- ❖ Company image perceptions

- ❖ Website usability testing
- ❖ Advertising testing

The uses of qualitative research are limited only by your imagination. Its main strengths are that it can be done quickly, it permits the customer to speak in their own vocabulary about your product and their expectations of it and it allows for surprise insights.

EXERCISES: Developing Qualitative Skills

1. Consider a competitive product or service situation, and develop a one-page discussion guide for a one-hour focus group. Give careful consideration to the order of the topics, and the length of discussion time for each topic. Will you be able to cover all the topics sufficiently in an hour? How will you guide the discussion to keep it on-track?

2. Using the discussion guide you developed above, hold an informal focus group with some business associates or friends in order to test out the group. Free pizza can often be a strong inducement to participate! Conduct the group using your discussion guide, while recording the session or taking notes.

3. Consider the information you gathered from your focus group, above. Can you organize the information you gathered into meaningful categories, and summarize the insights you gained from conducting the group? Was the group successful in providing you the information you were seeking? What might you do differently next time?

4. Try to attend a professionally-led focus group, to observe a trained focus group moderator in action. Some recruiters post focus group opportunities in the *Help Wanted* sections of local newspapers or on sites like **craigslist.org.** Participate in the group honestly and openly, but at the same time observe the techniques that the moderator uses stimulate, guide and control the group. Consider which of these you may want to use in the future.

9

Quantitative Market Research

Key Chapter Points:

- ❖ How quantitative market research differs from qualitative
- ❖ Introduction to quantitative research
- ❖ Designing and administering surveys
- ❖ Analyzing survey results
- ❖ Using surveys for CI
- ❖ Exercises to develop quantitative research skills

Quantitative Research

Quantitative or survey research differs from qualitative in both intent and execution. While the intent behind qualitative research is to explore subjects in depth with a fairly small number of people, quantitative is geared toward obtaining measurable results that can be projected to a larger population. To do this, you need to develop a survey questionnaire, with structured, mostly closed-ended

75

questions, and administer this questionnaire to a chosen sample of people.

A common misperception is that survey research is expensive. While it is true that a professional research firm may need to charge thousands of dollars to conduct a survey and prepare a report, it is equally possible for you to conduct your own surveys for virtually no cost.

The most important things to do correctly in survey research are:

- ❖ Accurately define the research objectives
- ❖ Develop an effective questionnaire that addresses the research objectives
- ❖ Choose the right sample of people to complete the questionnaire

If you follow these steps correctly, you'll be more likely to get accurate, error-free findings. What follows are the necessary tasks and processes involved in designing and conducting your own survey research and in making use of free web-based survey software.

Defining the Research Objectives

A lot of survey research goes wrong when people write survey questions without carefully considering what they want the survey to achieve. This practice of rushing ahead with questionnaire design almost always results in the wrong data being captured, data being missed, or data being gathered that does not do what it was intended to do.

The first step is to create a well-defined research objective. Think carefully about precisely what you want to learn from your survey. A weak research objective looks like this: "To figure out what our customers think of us and the other guys." This is a fine starting point, and may be in general terms what you are trying to find out, but it needs work.

You need to break this general idea down into its component parts until you arrive at very specific, measurable items. For example, you may want to:

- ❖ Measure how satisfied your customers are with your product quality, service and price
- ❖ Determine which competitors your customers also purchase from
- ❖ Uncover specific likes and dislikes about your competitors
- ❖ Profile survey respondents in terms of gender, age, education, occupation and local newspaper readership

Now the research objectives have been clarified substantially. When you have clear, measureable items, it is relatively easy to develop survey questions that will accomplish your objectives. If you had gone ahead with the earlier, vaguely-defined objective, you would be at a loss to translate it into meaningful questions. By spending the right amount of time and critical thinking on this important front-end activity, you will make the entire survey process easier, more efficient and more effective.

By defining your research objectives in sufficient detail, you can then design a questionnaire that will be as short as possible, with no unnecessary questions and that will be easy for respondents to complete. In general, most people are willing to take a short survey, particularly if it is clear and well-designed.

Online Surveys

There are a number of websites dedicated to helping you develop questionnaires on the Internet. Many of these sites are free. Examples are **Zoomerang.com** and **SurveyMonkey.com**. These sites offer good online guides to help you structure and develop questions and give you tips on how to develop the order of your questionnaire.

When developing a quick survey for competitive purposes, it can be extremely helpful to use a service like SurveyMonkey which allows you to quickly deliver your questions to a large number of people. Gathering intelligence faster and cheaper is what gives you a quick jump on your competition. This is one way to get the one percent lead that will help you stay ahead.

Student Resources

Many colleges and universities offer market research courses in which students are required to create market surveys for external companies and clients. This service is generally free or low-cost. Students can offer assistance in helping you understand your research objectives, in designing your questionnaire and in many cases they can also administer your questionnaire for you and write up reports with conclusions and recommendations.

To tap into this valuable resource, contact your local college to get the name of the market research instructor, and then give him or her a call.

EXERCISES: Developing Quantitative Skills

1. Consider a competitive product or service situation. Create a list of four to six research objectives that you want to explore, focusing on items that are measurable, and thus lend themselves to a quantitative survey. An example might be: "To determine whether the price of my product is seen as low, high or just about right."

2. Take the list of research objectives and transform them into survey questions. An example might be: "Would you say the price of (product X) is very high, somewhat high, about right, somewhat low or very low?"

3. When you are satisfied your questions will yield the information you want, go onto a free Web-based survey site like Survey Monkey. Insert your survey questions. (The *help* functions on the survey sites are very useful). When you're ready, publish the survey to the Web using the online instructions and send a link to the survey to several colleagues or friends.

4. When a few people have taken your survey, go into the results section of the survey website and consider the results. What are the results telling you? Are you getting the information you were looking for? What might you do differently next time?

10

Mystery Shopping

Key Chapter Points:

- ❖ Introduction to mystery shopping
- ❖ Structuring a mystery shopping operation
- ❖ Online mystery shopping
- ❖ Methods to practice mystery shopping

Mystery Shopping

In Chapters 1 and 2, you looked at using your powers of observation to gather CI. Taking a more systematic approach to these observation activities is called mystery shopping. Many firms do this internally on their own locations, but an equally valid approach is to include your competitors as well as yourself in the analysis. The key to mystery shopping is that it must be both systematic and consistent. To this end, a sound study design is essential, including a script (if necessary) and a rating form or scorecard of some sort. For a retail mystery shopping exercise, here are some logical dimensions to compare:

- ❖ Shelf location
- ❖ Sales rep expertise
- ❖ Up-selling and cross-selling by sales reps
- ❖ Pricing
- ❖ Length of time waiting in line-ups

When you analyze the mystery shopping results, you should be able to see the relative strengths and weakness between yourself and the competition. This will help you formulate counterattack strategies to reduce vulnerabilities.

Websites can also be a fruitful area for mystery shopping. You can compare your own website features and performance with that of the competition. To undertake this type of comparison, you need to first decide which are the most important or desired attributes of a website in your business. These can be a function of your goals and objectives for your web presence, and can also be determined by polling a cross-section of your website users. Some typical variables to measure include:

- ❖ Website load time
- ❖ Search engine positioning
- ❖ Number of clicks to purchase
- ❖ Ease of navigation
- ❖ Availability of live support
- ❖ Speed of response to email queries

Once you have isolated the key variables to focus on, you can conduct experiments on your website as well as your competitors' sites. You can measure items like load time with a stopwatch, and track other items using a set of structured activities. One good technique is to develop a common set of transactions to be done on each website.

Taking the example of a boat charter company, you can give your test users the following set of instructions:

1. Search for each company's site using Google and the keywords "boat rental"

2. Record the resulting position of each company in the search rankings

3. Record load time of each company's site in seconds

4. Try to sign up for a company email newsletter

    ❖ Does the first email newsletter arrive within one hour?

    ❖ Can you do this from the home page?

    ❖ How many clicks are required?

5. Go through the process of booking a sailboat

    ❖ How many clicks are required to see the boat layout and amenities?

    ❖ Can you book a boat charter online?

    ❖ How many clicks from the home page to booking?

    ❖ Can you get live support while on the site?

- ❖ Is the support telephone-based, online-based or both?
- ❖ Are privacy and security concerns addressed at the checkout?

6. Send an email query about the availability of a sailboat for this weekend. How long does it take to receive a response?

This is a partial list of features and performance items you might want to rate when comparing your website to your competition's. In order to remove any guesswork in deciding what to rate, you could convene one or two focus groups with existing and prospective customers to talk about:

- ❖ How they use your website
- ❖ What their expectations are of services and performance
- ❖ What they like and dislike about the various websites

Once you have isolated the most important features and performance variables, you can develop the script for your testing, as in the boat example above. The results of your research can be easily assembled into a CI matrix like this:

	Driver 1	Driver 2	Driver 3	Driver 4	Threat Level	Counterattack strategy
Your Company						
Competitor 1						
Competitor 2						

This allows you to assess which competitor's website presents the greatest threat, and helps you formulate a counterattack response.

This online mystery shopping exercise can be repeated as often as you need depending on the how dynamic your competition is. Given that it is relatively easy for most firms to make adjustments to their web presence, online mystery shopping is one of the most potent and cost-effective forms of CI.

Casual and Opportunistic Mystery Shopping

Formal mystery shopping techniques benefit greatly from being well-designed and systematically-executed. These attributes contribute to the robustness and verifiability of the intelligence you'll gather. However, this does not mean that all mystery shopping has to be formal and highly-structured to result in useful intelligence.

It is a good habit to be continually mystery shopping, so that it becomes second nature. For example, if you own a chain of coffee shops, you can visit other coffee shops, consume the product, check products and prices, and observe service levels. Similarly, when you're in other regions or countries, check out comparable outlets, possibly uncovering a new trend to incorporate back home before your competition.

EXERCISES: Developing Mystery Shopping Skills:

1. Choose a service industry in your local community that interests you. Using the ideas in this chapter, create a set of service dimensions that you would like to measure in your competition. These might include factors such as:

    ❖ Whether the salesperson acknowledged you when you entered the store
    ❖ Length of time spent waiting in line
    ❖ Quality of the food or products
    ❖ Pricing
    ❖ Helpfulness of staff
    ❖ Courtesy of staff
    ❖ Knowledgeability of staff
    ❖ Cross selling or upselling
    ❖ Any other features that you think are useful to know

    Remember: The key to mystery shopping is to keep your criteria as objective and repeatable as possible. For example, it is difficult to objectively measure courtesy. However, you can measure whether somebody smiled or made eye contact when you entered. By making your measures as specific as possible, you will get more effective results. If possible, repeat this exercise at other outlets of the same business. Then, write up your findings in the form of the CI Matrix.

2. This exercise will help you practice casual mystery shopping which will sharpen your observation skills. For a single day, every time you're in a situation where there is a store line up, practice counting how many people are in that line-up and how long it takes to get the cashier. At the end of the day, work out some conclusions that you've drawn from your observations. Which businesses move their line up more quickly? Why do you think that is? Ask and answer a few questions like this using skills you developed in earlier chapters.

11

Human Intelligence

Key Chapter Points:

- ❖ Introduction to human intelligence-gathering
- ❖ Active listening
- ❖ Verbal and non-verbal communications
- ❖ Elicitation techniques
- ❖ Planning and structuring human intelligence operations
- ❖ Methods to practice human intelligence-gathering

The Power of Human Intelligence

Human sources of intelligence are among the most powerful ways of gaining a strategic advantage. The human mind has an impressive ability to combine facts and knowledge to create insights. As you gain experience with CI, you'll be impressed by the ease with which you can extract these insights. In this chapter, you'll explore the most powerful tools for listening and speaking to people, and getting them to volunteer intelligence. Nothing presented is unethical, and the techniques are easy to learn and practice.

Active Listening

Active listening is a tool practiced by counselors, focus group moderators, salespeople and other professionals in order to encourage others to talk. It involves focusing on listening rather than speaking. The main techniques involve:

- Listening intently, with all your focus
- Avoiding the temptation to inject your own opinions
- Offering non-verbal encouragement through nodding, smiling, and murmured acknowledgements
- Mirroring the other person's body language

Active listening can yield impressive results, since many people feel as though they are not listened to as much as they would like. Active listening is a tool that can really open the floodgates of human intelligence.

Non-verbal Communication

Studies have shown that people communicate chiefly through body language and tone of voice, as opposed to the actual words they say. One measure suggests that communication is 55% body language, 38% tone of voice, and only 7% the actual words themselves. This means that you risk ignoring 93% of a person's meaning if you only focus on the words they are using. If you don't train yourself to be alert to body language and tone of voice, you'll lose a valuable amount of intelligence.

Focus group moderators in particular are trained to take into account all aspects of a respondent's communication. This is why most focus groups are videotaped, so that the body language and tonality of the respondents can be analyzed and considered later.

Some body language cues that you should train yourself to be alert to include:

- ❖ Overall body posture: Tense or relaxed? Open or closed?
- ❖ Arm position: Crossed and guarded?
- ❖ Eye contact: Maintained or avoided?

In studying a person's body language, you need to be observant to how it changes over time and in response to particular topics.

Tonality or tone of voice offers a significant source of insight. By paying particular attention to tonality, you can sense changes in the meaning of what people are telling you. Things to listen for include:

- ❖ Anger
- ❖ Tension
- ❖ Nervousness
- ❖ Happiness

- ❖ Deception
- ❖ Vindictiveness
- ❖ Criticism
- ❖ Cynicism

As is true for body language, you need to train yourself to listen for changes over time and in response to different topics. When you are trying to elicit intelligence from people, an increase in the comfort level of the subject's tonality can signal that it is safe to probe more deeply or to steer the conversation in a direction that you want.

Taken together, active listening and attention to tonality and body language provide a powerful human intelligence-gathering toolkit. By employing these techniques, you can learn far more from human sources than your competitors can. In the balance of this chapter, you will round out your human intelligence toolkit by learning how to combine these techniques with a set of tools called elicitation.

Elicitation

Elicitation, in its simplest definition, involves getting people to tell you things voluntarily and without being asked directly. The skills presented here are derived from the work of law enforcement and intelligence agencies. Let's focus on a few key techniques of eliciting information from human sources. The first involves a way of structuring a conversation to allow for the effective use of elicitation techniques.

The Conversational Hourglass

The ideal outcome for a human intelligence interaction is that the target volunteers the intelligence you desire without being asked for it, and without particularly remembering giving it to you. You can often achieve this by designing the conversational flow using the method described below.

The theory of the conversational hourglass is that people tend to remember:

- ❖ The beginning of the conversation
- ❖ The end of the conversation
- ❖ Direct questions, as opposed to indirect or elicitation

What happens in the middle of conversations tends to be most easily forgotten, and it is in this "zone" that you want to use elicitation techniques. The conversational hourglass refers to a design of a conversation that begins with direct questions and ends with direct questions, usually on topics unrelated to the area of intelligence-gathering.

Elicitation Techniques

Remember: Elicitation means getting people to offer up information without asking directly for it. Before getting into specific techniques, it is useful to consider why people talk at all. Some reasons include the following very basic human traits:

- ❖ Need to inform or teach
- ❖ Need to feel listened to
- ❖ Need to correct others
- ❖ Desire to complain
- ❖ Desire to seem like an expert
- ❖ Tendency to lose control when emotional or "under the influence"

There are a number of elicitation techniques used by intelligence professionals to take advantage of these character traits; what follows are a few of the most useful.

- ❖ **Active listening** is one of the most powerful elicitation skills. By simply mirroring body language, listening intently, and using verbal and non-verbal encouragements, you can keep your subject talking, only interjecting gently from time to time to nudge the conversation along certain directions.

- ❖ **Playing dumb** is also a powerful elicitation tool. If you appear to be an expert in the field you are trying to learn about, you are likely to trigger a competitive or protective reaction in your subject. People are more likely to be guarded if they think that the person they are talking with understands the value of the intelligence being discussed. By credibly playing down your level of knowledge, you will encourage more open discussion, particularly as far as details and specifics are concerned.

- **Faking disbelief** or acting as though you don't believe what you are being told can cause the subject's need to correct or instruct to kick in; they will volunteer more and more proof to back up their assertion. This disbelief doesn't necessarily have to be verbal, since non-verbal cues and expressions may be enough to keep the subject talking.

- **Provocative or erroneous statements** are designed to get a rise out of someone. If you say to a person that you have heard that their new product launch has been set back several months, they may correct this misinformation with a more accurate launch date – the piece of information you were seeking.

- **Oblique references** can be a steering device for a conversation. By referring to a topic that is conceptually related to what you want to uncover, you can steer the respondent toward the topic you want to learn about.

For example, if you want to determine if a competitor is downsizing their finance department, you can begin by mentioning that your own human resource department is downsizing. This plants the seed of discussing downsizing in the subject's mind without actually bringing up the specifics of what you are looking for. Used in combination with other tools like word repetition, you can bring the subject around to volunteering the information you are seeking.

- ❖ **Word Repetition** is another device for steering a conversation around to a topic you want to discuss. This technique involves us repeating the same word or concept, worked into your side of the discussion. For example, if you want to learn about your competitor's plans to downsize their finance department, you can mention terms like "finance", "number-crunchers", "accounting types" repeatedly to plant the seed of discussion in the subject's mind.

- ❖ **Flattery** is a surefire key to success. If you make someone feel good about something they are doing through praise, they will be encouraged to keep talking about it. If you flatter the subject by quoting favorable press articles and the like, they will be likely to keep revealing more about their firm's successes.

- ❖ **Complaining:** We are all prone to it if given the chance. By complaining about an aspect of your own current situation, you can often encourage the subject to open up with some complaints about their own experience. For example, if you complain about how your company is always over-promising and under-delivering on new product launches, which makes life difficult for your sales reps, your subject may weigh in with some complaints of their own, allowing you to uncover vulnerability in your rival.

- **Silence** is a well-known focus group moderator's tool. By not speaking, you encourage the other person to fill in the silence. This can be useful to keep the person talking. This tool is helpful when you want the subject to speak in more detail or depth about the topic at-hand.

Tying It All Together

The conversational hourglass approach involves tying all these elicitation techniques into an overall strategy for extracting a particular piece of intelligence from a source. Start by determining what specific piece or pieces of intelligence you want to extract. Once you know what you are seeking in very concrete terms, you can design a conversational strategy. Recall that people tend to remember the beginnings and endings of conversations, as well as direct questions. So, you will pepper the beginning and end of the conversation with direct questions, and use elicitation in the middle, hence the hourglass image. A plan for an hourglass conversation using elicitation might look like this:

Desired piece of intelligence: When is my competitor releasing product X?

Target: Bill Smith, Manager of Research & Development

Location: Trade show and conference, 2nd day

Introduction:

- ❖ Are you a hockey fan? (you know from his website bio he is)
- ❖ Did you go the game last night?
- ❖ So who do think is going to win the Cup?

Elicitation:

- ❖ I get sick of being sent to these things (complaint)
- ❖ I'm supposed to be telling everyone that our product is just about to hit the market, but I know it is delayed as usual (complaint)
- ❖ You guys seem to hit the mark every time though (flattery + erroneous statement)

Closing:

- ❖ Are you going to able to catch a game while you're here?
- ❖ And you really think the Bruins are going to win?
- ❖ It's been good chatting with you.

The above example is simplified, but it shows the hourglass structure of starting and ending with direct questions, using only elicitation in the middle. When the target walks away from the conversation, he will remember the hockey conversation because it consisted of direct questions that happened at the beginning and end of the conversation. The middle of the conversation will be a blur.

An important thing to remember with human intelligence-gathering is that you need to remain flexible. It helps immensely to have carefully defined what specific intelligence you are looking for, and to have sketched out an hourglass design for the conversation, with some specific techniques in mind. However, conversations have lives of their own, and it is better to go with the flow rather than keep trying awkwardly to bring the conversation back to where you want it to go. It is usually better to back away from the encounter and try again another time or with another person. The worst thing is to be seen as snooping, because word will get around and opportunities will dry up quickly.

Choosing Respondents

As mentioned earlier, conferences and trade shows can be an excellent source of human intelligence targets. You can usually determine who is speaking at the conference as well as find their company bio and their photograph by looking at the conference program. In addition, you can usually see where each exhibitor's booth is on a floor map, so you can plan where to "bump into" people.

Cross-cultural Communication

Much of the human intelligence material you have considered here relates to American and Canadian business cultures. Clearly every culture around the world will have its own ways of doing things and social cues. Even different states and provinces in North America will differ in some significant ways.

It is obviously prudent then to secure some local knowledge of a new culture you may be thinking of entering to gather intelligence. Hiring a local CI expert as either a coach or an operative may make sense. At a minimum, you should do some Internet research to learn how communication styles may differ.

Style Flexing

In this chapter, you have learned a number of tools for better communication, and ultimately better extraction of human intelligence. One final element is the importance of style flexing, a skill taught in sales programs. This involves bending your communication style so that it more closely resembles the other person's. This dramatically enhances conversational success and information flow by building rapport between the two parties.

EXERCISES: Gathering Human Intelligence

1. Developing active listening skills: Take an everyday conversational situation and practice active listening for five minutes. After a topic of discussion has been established, encourage the other person to talk openly through listening intently, mirroring body language, using verbal encouragements such as repeating words, and making use of non-verbal encouragements such as "mm hmm".

   Make a note of what worked well and what you would improve next time. Practice active listening a few times a week, and it will soon become second nature.

2. Observe non-verbal cues: In an office, café or other public setting, discreetly observe someone nearby who is engaged in a conversation. What meta-messages can you infer from his or her eye contact, posture, volume and pace of speech, hand movements and so on?

3. Planning a conversational hourglass: Before you enter into a social situation, choose a piece of information you would like to elicit from someone. Plan a strategy using the conversational hourglass. How will you begin the conversation? Which elicitation techniques might work best given what you know of the situation and the person? How will you close the conversation? After the conversation is finished, take note of what worked well and what you can improve next time. Practice this on a daily basis until you are comfortable with it.

4. Research cross-cultural communication: If you frequently interact with someone who is from a culture different from your own, do some research into that culture's verbal and non-verbal communication styles. Then try style-flexing, bending your own communication style to meet that of the person who comes from a different culture.

12

Trade Show Intelligence

Key Chapter Points:

- ❖ Introduction to trade show intelligence
- ❖ Types of intelligence available at trade shows
- ❖ Planning and structuring trade show intelligence missions
- ❖ Methods to practice trade show intelligence skills

Trade Show Intelligence

Trade shows are an excellent source of intelligence, both human and written. Attending a trade show with a calculated view to obtaining CI can be a very cost-effective exercise. The key to trade show intelligence is first to determine your intelligence objectives. Once you have decided what specific pieces of intelligence you desire, you can formulate a plan of attack, identifying likely sources of CI as well as who will be responsible for obtaining each piece of intelligence.

You can plan much of your CI mission for trade shows in advance using the Internet. Most trade shows have websites that can be mined for useful information such as who will be speaking. As well, there is usually a map layout of where each competitor will be located. You can often find photographs of specific people you may wish to approach for intelligence.

In approaching a tradeshow for intelligence-gathering purposes, team organization is very important. There are several different roles to fulfill:

- **Team leader:** Responsible for planning, organizing, setting intelligence goals, determining specific targets, assigning each intelligence objective to field operatives, communicating final results

- **Field operatives:** Responsible for obtaining specific pieces of intelligence, through physical collection, observation and human intelligence

- **Analysts:** Responsible for pre-trade show intelligence as well as gathering, collating and analyzing intelligence gathered from field operatives

The reality for many is that you are a "CI department of one," and will have to perform all of these functions alone. The alternative may be to hire students from a school that teaches market research and CI; they are often thankful for paid entrance to a useful trade show. In any event, careful consideration of each of these functions is necessary in order to carry out a successful trade show intelligence-gathering mission.

The role of the team leader is first to plan the exercise, which involves determining what intelligence is desired and which trade show it makes sense to target for the intelligence. Once a trade show has been selected, online research and/or printed marketing material is helpful in terms of refining and selecting your targets for the intelligence you require. Often such specifics as lists of exhibitors, floor layouts, speaker biographies and the like are available. This is useful because it helps avoid unnecessary wandering during the trade show itself.

The next task is to assign specific pieces of intelligence to the person responsible for gathering it. Each piece of required intelligence needs to be defined in terms of how you will obtain it. In the case of human intelligence, you need to determine who you will target and what approach you will use. The conversational hourglass approach to elicitation covered in the previous chapter will be useful here. For other intelligence, other approaches will make sense. Perhaps you want to get a look at the features of your competitor's soon-to-be-released product. Then you'll try to see if printed material or demonstrations are available.

One of the exploitable vulnerabilities companies face at trade shows is that it's usually enthusiasts who are selected to staff the booths. These can be employees who are too inexperienced to know what should be kept secret, or sales people who are eager to promote the products to anyone who might be interested, or product development experts who can be lured into detailed discussions of the features that they are proud of.

At the trade show itself, particularly multi-day events, you should begin by doing a sweep of the facility, gathering everything that is available in terms of printed materials and giveaways from all the relevant target companies. This "snag and bag" operation can often yield plenty of the intelligence that you want. Taking the gathered material away and analyzing it can help you to refine your approach and targets for the next wave of intelligence-gathering. At an out-of-town trade show, your hotel room can serve as a command post, allowing you to sift through the material you've gathered.

Then, you can refine your plan based on what you extract from the material.

In general, the less face-to-face interaction you rely on to obtain your intelligence, the better. Anything you can gather by indirect means like printed material, observation or eavesdropping is a bonus, since it makes it far less likely that your intentions and activities will be detected. This multi-wave approach to gathering intelligence allows us to use the riskier tactics, like human intelligence sparingly, thus minimizing the chances of being seen as "snooping."

A three-wave approach to a trade show intelligence-gathering mission can be summarized as follows:

- ❖ **Plan the overall mission:** Determine the required pieces of intelligence and the targets or sources for the intelligence

- ❖ **Wave one:** Make a first pass of the trade show, gathering everything that is being given away at the target booths. Analyze the gathered material and see if some of the required intelligence is there, then refine your requirements

- ❖ **Wave two:** Conduct human and observational intelligence, then analyze this intelligence to isolate any remaining gaps

- ❖ **Wave three:** Conduct final intelligence to gather any remaining information you require

EXERCISE: Trade Show Intelligence Planning

1. Using Google or similar Internet search tool, find a trade show that is relevant to your business, one in which you might expect to find some useful competitive intelligence. Locate the website for the trade show. Try to find out who is attending and who will be speaking at the show. Pick one target from whom you would like to gather human intelligence.

   ❖ Can you determine what that person looks like from the conference website or other resource such as Google Images, so that you would recognize them at the show?

   ❖ Can you determine where they are likely to be on a given day from the conference speaking schedule?

   ❖ Can you learn more about their background, hobbies and interests from other sources?

   Work up a conversational plan to elicit information from that person, incorporating the human intelligence techniques from Chapter 11.

13

Executive Profiling

Key Chapter Points:

- ❖ Introduction to executive profiling
- ❖ The role of biographical sketches and personality types
- ❖ Remote personality typing
- ❖ CI and other uses of executive profiling
- ❖ Methods to practice executive profiling

Executive Profiling

Developing a profile of the key executives in rival firms can be an invaluable source of predictive intelligence. If you know how someone has made decisions in the past, you can predict how they will make decisions in the future. For example, if you lowered your price, how aggressively would they respond? This chapter presents a simple and inexpensive set of techniques for profiling key executives.

You will develop an executive profile based on two main components. The first is a biographical sketch, and the second is a simple personality profiling test. The theory behind this approach, developed by Dr. Marta Weber, is that you can predict how a person will react to situations by knowing how they have behaved in the past. You can also uncover their basic personality type or problem-solving style. What follows is a simple approach to these two components.

In brief, the profile involves combining the results of a biographical sketch with the results of a personality test "guesstimate" of the profiled executive. Taken together, these two components give us a potent predictor of how the target executive thinks, approaches problems, and reacts to new situations.

Biographical Sketch

For a biographical sketch, the Internet is a logical starting point. By Googling a person by name, you can often uncover articles, biographical profiles, volunteer affiliations and many more useful facts. The better known the person is, the more you can find out. However, even local people with a smaller profile often have personal websites and other useful sources of personal information. Here are some things to look for:

- ❖ Age
- ❖ Place of birth
- ❖ Poor or rich
- ❖ Family situation

- ❖ Schools attended
- ❖ Religious and cultural influences
- ❖ Post-secondary education: What, how much, where
- ❖ Early career positions
- ❖ Mentors
- ❖ Important business decisions
- ❖ Current family situation
- ❖ Hobbies
- ❖ Lifestyle

EXERCISES: Developing Profiling Skills

1. Pick a famous executive, and use the Internet to develop a biographical sketch, trying to find as many of the above elements as possible, as well as other interesting facts that you uncover. Now, sift through the material and try to come up with a short capsule summary of their make-up. How would he or she respond to a situation like a hostile takeover attempt? Run? Fight back aggressively? Cooperate, trying to extract the best price for shareholders?

2. There are a number of personality tests available to help you complete a personality profile. Consider using the Myers-Briggs personality test, versions of which are available for free by Googling Myers-Briggs. (One link to a simplified version of Myers-Briggs can be found at **humanmetrics.com**.)

   The Myers-Briggs Type Indicator or, MBTI, measures people along the following dimensions.

   - Introversion versus extraversion *(roughly, whether you are energized or drained by social interaction)*

   - Intuition versus sensing *(roughly, whether you pay more attention to your intuition or your five senses)*

   - Thinking versus feeling *(roughly whether you pay more attention to logic or emotion)*

❖ Perceiving versus judging *(roughly, whether you prefer to flow and adapt or to have routines and structures)*

The result of the test is a four-letter type, such as INTJ, that you can assign to a person to help you understand how they approach people and situations.

To determine both your MBTI type and that of another person, go to **humanmetrics.com** and take the survey. Next, drill down into the description for your type and see how your type approaches situations. Does this ring true for you?

3. Now, attempt an MBTI profile on another person, perhaps the same person you did the biographical sketch on. There are two ways to do this – one is to re-take the online test as if you were the other person, similar to how an actor gets into a character. The other, more shorthand way would be to assess each MBTI dimension in light of what you know about the person. For example, are they an introvert or an extravert? If their hobbies are reading and quiet times at the lake, you might guess they are an "I", whereas a pattern of much public speaking, hosting parties and an active interest in performing arts might tip us off to them being an "E" or extrovert.

Carry on typing the person until you are reasonably certain of their MBTI type. Now, go into the explanation of the person's type and consider how they approach people, information and problems. Can you use this insight to find exploitable vulnerabilities in their character make-up?

4. You have developed a biographical sketch of a person as well as a guess as to their MBTI type. Now, try creating a complete profile. Consider these two elements together, in order to look for exploitable weaknesses. What is the best way to compete against someone who shows a tendency toward egomania and boasting in their biographical sketch? Is it to stage a public bragging contest through press releases, and draw them into a fight? If this person is an intuitive (N) as opposed to a sensor (S), is the key to winning this bragging war going to be drowning them in data? For the person you have profiled, consider their strong points and vulnerabilities, and how their weakness could be played upon and exploited in various situations. This will help inform your counterattack strategies in the future.

Entrepreneurs can use the above tactics, combining a biographical sketch with a remote personality typing, to assist in selling key accounts. The same information can be used in a cooperative, solution-oriented scenario such as sales or a job interview. By developing the deepest possible profile of a person before you approach them, you can better craft a successful approach. You will know better how to communicate as well as which particular interests you should play off.

PART THREE:
Tying It All Together

A Set of Tools

CI involves a set of tools, tactics and techniques that can be learned and generally employed at little or no cost. Many of the tools presented in this book can be done from your desktop, and many others require only a small amount of personal, physical and observational effort. Some of the tools you learned in this book include observation, analysis of Internet newsgroup chatter, mystery shopping, focus groups, surveys, human intelligence, website benchmarking and executive personality profiling. The tools themselves, while powerful, are not complex. Your challenge is to employ them in a strategic manner, creating a whole CI system geared toward deriving sustainable strategic advantage.

A Way of Being

In this way, CI is really a way of being, not a guise or function that is confined to the work day. To be really successful, you need to internalize the CI skills so you are seeing and acquiring intelligence all the time, on the job or off. Success in CI requires two key attributes: An ability to see intelligence everywhere and a ceaseless desire to acquire and employ intelligence.

14

CI as a Daily Practice

In Part One of this book, you learned to develop or practice skills to become a better observer, hone your ability to gather information by using your five senses, improve your numeracy and develop your knowledge of ethical practices.

Part Two helped you develop applied CI skills in everything from industry analysis to human intelligence to executive profiling. Now you have a complete toolkit for practicing competitive intelligence quickly, cheaply and ethically.

It's essential that from this point forward that you begin to assimilate these CI skills into your daily life. Competitive intelligence only succeeds if it becomes an internalized practice that you do almost subconsciously. No amount of intelligence can provide you with a competitive edge if it is something you gather only once or twice, or if your results end up gathering dust in a binder on a shelf.

The following key tools will keep your CI system working for you on a regular basis. First is the CI matrix reporting system. When done correctly, with insight into the drivers of competitive success in your industry, the matrix provides you with a powerful yet simple one-page tool to help you:

- ❖ Decide which items you need competitive information on
- ❖ Summarize those items in an easy to read and understand format
- ❖ Clearly define tactics to help you develop counterattack strategies
- ❖ Stay ahead of your competition.

The CI matrix reporting system is a powerful tool because it is:

- ❖ Simple
- ❖ Focused
- ❖ Easy to update and distribute
- ❖ Scalable

Vital Steps for a Vital System

By following the steps below, you will increase the value of your CI efforts, without expending excess time, cost and energy.

1. Keep the system simple. You're much more likely to abandon any process too complicated to maintain and upkeep.

2. Keep the tactics and techniques that drive the CI system as repeatable and as well-defined as possible. For example, when mystery shopping, make sure you create a form that measures and records the same characteristics the same way each time you use it. This allows the form to be used by different people at different times while standardizing the results.

3. Keep your system current. Information that isn't updated quickly becomes irrelevant; this is where your competitors will gain back the advantage that used to be yours.

4. Keep your reporting system relevant by focusing on the key drivers, the ones that are critical to your business success. If you focus too much on "nice to have" information rather than the essentials, you risk wasting time studying information that's not really important.

The CI matrix reporting structure is very versatile. It can be adapted for many situations including overall strategic assessments, mystery shopping and competitive website comparisons. It's possible to have a number of matrices that measure different aspects of your business including different departments, products or services. The matrix's key strength is its ability to summarize a lot of information, concentrating the reader's attention on the most essential elements driving competitive success. Although the size and makeup of the matrix can be adjusted endlessly, keep the following guidelines keeping in mind:

- ❖ Keep it to one page
- ❖ Focus only on the key drivers of competition

❖ Focus only on the most troublesome competitors

The threat level of your competitors should be tracked and can be measured from one to ten. The threat level can be calculated mathematically but it's faster to and possibly just as useful to eyeball who is winning competitively. Remember to keep your counterattack strategies realistic and achievable and to update the matrix as often as is dictated by the dynamics of the environment. It's not a static document.

How and When to Gather CI

Keeping your CI system working for you also involves the process around how and when you gather information. First, spend an appropriate amount of time deciding what intelligence is essential to gather. This can be done using the industry analysis techniques and the CI matrix.

Next, consider what you need to know. At least 30% of your competitive intelligence effort should be spent on deciding what information to gather. You can greatly increase the efficiency of your data gathering efforts by excluding 70% of the information—the facts and figures that are irrelevant to your competitive success.

Keep your networks warm. You can gather a vast amount of human intelligence either in person or by on the telephone. But remember: Don't only interact with people when you need something from them. It's far better to do favors for people who may be in a position to give you intelligence at a later date. Offer them a little intelligence of your own in return. These *quid pro quo* relationships help keep your networks warm by encouraging others to maintain a friendly disposition toward you, so they'll be there when you need them the most.

Assign responsibilities to particular people in your organization. Realistically, for the entrepreneur, it may be that all the responsibilities are yours alone. However if you do have a team of analysts or assistants, remember to be very clear in assigning who does what, who is responsible for monitoring the pulse of particular areas, who needs to track what information. As well, it's important to find a tool – perhaps the CI matrix—that communicates the findings of your intelligence in a cogent, focused and actionable manner.

If you are in a larger organization, it's important to spend an appropriate amount of time lobbying senior management regarding the importance of competitive intelligence. Although CI has been included in business school curricula for several years now, many senior executives aren't aware of its uses. Traditionally, CI has been seen as something the market research department does "somewhere downstairs." Only now are senior executives starting to realize how critical it is to have good intelligence. However, as new generation of CEOs begins to emerge, CI is rising in prominence.

Looking at the Past, Present and Future

Every time you do a full CI audit of your company and its competitors, you go into it with certain assumptions about what is important to the industry, who are the most important competitors and what are the most important drivers of competitive success.

In order to keep your CI system alive and vibrant, it's important to take some time each year (or even more frequently if you are in a particularly turbulent industry) to revisit various aspects of your intelligence program.

First, revisit the models on which you built your original intelligence. Test the assumptions you used both the first and last times you gathered intelligence. Are they still valid? Similarly, you will want to assess which competitors you should focus on now as opposed to when you last did an assessment or analysis.

Keep in mind that the drivers of competitive success sometimes change. If you are using scenarios, it's important to revisit them as current events change the landscape of your industry. For example, you may have been speculating that two major competitors would merge. If they did merge, it's important to think about what could happen next. As well, take current events into account. How will oil price changes, industry regulations, or other factors affect your situation?

One way to update your system is to have a CI retreat once a year in which the key intelligence personnel in your organization brainstorms together. This allows you to test, retest and evaluate the different scenarios and ideas related to the current status of your competitive landscape.

During these sessions, it's important to move from a past or present orientation to a future orientation. This is known as predictive competitive intelligence. Predictive

CI tries to assess what might possibly happen and what will probably happen in the future. There are a number of ways to make these predictions. One of the best is to organize a predictive CI brainstorming session to examine factors such as:

- The model of your industry
- The nature of your competitors
- The nature of the executives who run the companies

You can use these factors to develop a wide variety of scenarios. For example, it's worthwhile to look at the most likely scenarios faced by the companies in your industry such as mergers. Use brainstorming to figure out what early warning signals will alert you to a merger. Perhaps you hear an investment bank has been hired by one of your competitors. This would be a key signal or "trip wire" to alert you to the fact that something very tangible is happening, probably involving a merger or acquisition.

Another tool that can be useful is a simulation or wargaming. If you have enough people in your company, divide into two different teams. One team plays your company; the other team plays your competitor. Use these two teams to do a mock battle with each other, putting them through different scenarios that could happen. For example, what if the government regulated your industry in a certain way? What if social trends made your customers' purchasing habits change?

A useful exercise is to imagine the least likely but most catastrophic action your competitor could take. An example of this in terms of national intelligence was the

9/11 attacks. An extremely unlikely, extremely destructive action is the one most likely to catch a company off guard. Have your teams brainstorm and simulate a disaster scenario. An example of this would be the banning of trans-fats in the food industry: The most able competitors would already have a back-up plan for this situation.

By testing the two competing teams in a variety of potential developments, you can come up with a deeper understanding of how your competitors may react in response to industry dynamics in the future. The more you know about what might happen in the future, the more you will be to anticipate and counteract what your competitors could do.

Conclusion

Using the skills that you've developed, you can now create a vibrant CI system, one that is constantly being challenged, updated and refocused and one that is sure to lead you toward competitive success. The skills you have picked up in the preceding chapters are ones that do not require massive investments of time, money or effort. In fact, as you practice these skills, many of them will become second-nature, like gathering intelligence at trade shows, using human intelligence skills to gather information from people, and profiling people before you meet them. Most importantly, all these techniques are ethical.

Now get out there and compete... intelligently!

About the Authors

Photo by Scott McAlpine

Rob Duncan is the Director of the Applied Research Liaison Office at BCIT. In his day-to-day professional work, Rob employs a variety of marketing intelligence techniques to assist entrepreneurs in getting new and improved products into the marketplace. Rob has also spent many years as a consultant and college instructor in the marketing and competitive intelligence fields. Rob holds a BA in Economics, an MBA, and is a Certified Management Consultant and Certified Marketing Research Professional.

Rob Duncan is also the author of *Haul Away! Teambuilding Lessons from a Voyage around Cape Horn*. The book shares teambuilding and management lessons learned from Rob's 75-day voyage around Cape Horn on a tall ship. When he is not at the college or consulting, Rob can often be found on his sailboat in the Pacific Northwest.

Alex Tyakoff is a senior intelligence analyst with Delta Police in British Columbia, Canada and a faculty member with the BC Institute of Technology (BCIT) where he teaches intelligence analysis. He is adjunct faculty with California State University (San Bernardino), National Security Studies.

Prior to his current position at Delta Police, Alex Tyakoff served as a tactical analyst with the Organized Crime Agency of BC. He was employed as a policy analyst with the Police Services Division and served as a strategic analyst with the Coordinated Law Enforcement Unit.

Mr. Tyakoff holds a BA in Political Science and a MA in Planning. His research interests include data-mining, systems dynamics and modeling, and social network analysis.

Tracy Urban is a Vancouver-based writer and researcher. She holds a BA in English and a MA in Curriculum, Teaching and Learning and is the owner of StoryHeart Productions, (**storyheart.com**) a company that specializes in helping professionals get their ideas into print.

Printed in the United States
117274LV00001B/55-162/P